Breeding Poultry For Exhibition

by E.F.Z.S. Cobb

with an introduction by Jackson Chambers

Self Reliance Books

Get more historic titles on animal and stock breeding, gardening and old fashioned skills by visiting us at:

Introduction

I am pleased to present yet another title on Poultry.

The work is in the Public Domain and is re-printed here in accordance with Federal Laws.

As with all reprinted books of this age that are intended to perfectly reproduce the original edition, considerable pains and effort had to be undertaken to correct fading and sometimes outright damage to existing proofs of this title. At times, this task is quite monumental, requiring an almost total "rebuilding" of some pages from digital proofs of multiple copies. Despite this, imperfections still sometimes exist in the final proof and may detract from the visual appearance of the text.

I hope you enjoy reading this book as much as I enjoyed making it available to readers again.

Jackson Chambers

BREEDING POULTRY

FOR

EXHIBITION.

F W

SECTION VII.

To the scientific poultry breeder—the man who has, to the best of his abilities, studied the cause and effect of hundreds of matings—nothing connected with poultry would seem more easy to write about than the above subject—anyhow, as far as those breeds that he personally had bred successfully. As a matter of fact there is no difficulty for such a one. He has all the points of mating at his fingers' ends, and he could at any moment sit down and write off all he knew about the mating of any breed or breeds that he is thoroughly acquainted with, and be able to follow out those instructions successfully at any time he was called upon to do so.

No, there is no difficulty about that. The difficulty is, could anybody else understand those instructions?

Have you ever listened to half-a-dozen expert fanciers discussing the shade of colour, or barring, or pencilling of any given breed? If so, you will also have heard many such questions as these: "But what do you mean by a rich buff?" "What is your idea of a heavily-barred bird?" "How wide would the pencilling be on what you term 'a finely pencilled bird'?" and so on and so on right throughout their conversation.

If, therefore, experts talking together are unable to properly understand each other without a number of questions being asked and answered, how much more difficult is it for a writer to instruct a novice whom he never sees.

For this reason we have endeavoured as far as possible to confine our remarks entirely to the ideal show specimen for the time being. We have endeavoured throughout to always have the ideal show bird in our mind, and not to write as if we ourselves were giving our personal definition. For example, "a too heavily barred bird" means not a bird with bars of a certain width, but simply that they are too wide for the ideal show specimen at such time as you happen to be mating up your pens.

Always bear in mind that a popular exhibition breed or popular exhibition variety of a breed is continually changing in the ideal. That is to say, there are few standards drawn up by the club of any particular breed that are not altered occasionally; or, if they are not, one of two things has happened— viz., either the winning specimens of the day are ahead of—in the eyes of judges and fanciers—the definition as stated in the club standard, or else the breed has come to the height of its career as a show specimen, and, unless fresh impetus is quickly given, will not merely remain stationary, but very soon drop in the scale of public favour.

A real fancier is, consciously or unconsciously, always striving for something better than he or anyone else has ever produced before. If he produces the ideal bird that he had in his mind's eye ten years ago—which quite possibly was in advance of the club standard—he has most likely by that time conceived an ideal for the breed far in advance of his former conception, and so still goes on striving for something better.

We trust, therefore, the reader will bear in mind, because this is important, that throughout this section of our book we are striving to show him how to breed any point he may either now or in the future desire to, and not merely to instruct him how to breed a winner of to-day, but to accomplish his object even though the Club standard is totally different from what it is at the time of writing.

COMMENCING OPERATIONS.

We would strongly advise anyone desirous of entering the Fancy, or anyone already in the Fancy, and who thinks of going in for another breed, not to do so until he has made himself thoroughly acquainted with the points necessary for a good specimen of that breed. A good plan to acquire this knowledge is by attending several shows, and requesting the judge or some fancier present to kindly point out to you why the first prize bird is placed ahead of either the second or third bird. A few hours spent in this manner will teach you more of what is required in a good exhibition specimen than several years careful study and breeding in your own yard.

The chief difficulty that presents itself to the follower of this course is to know whether the information imparted to him can thoroughly be relied on. We do not wish to imply that either judges or fanciers in general are any worse or any better than the average person one meets. But it is only natural—unless the judge happens to see he has made a mistake (which is rare) and is willing to own it (which is more rare)—that he should endeavour by argument to prove to you that the first prize bird is ahead of the second prize bird, and that the third prize bird is out of it altogether. But, if instead of tackling the judge you fall in with the disappointed exhibitor your case is almost hopeless. His goose is a swan, and nothing that you or the judge can say will ever alter that fact. "Didn't my bird win first at the Wasters' Show? Well! what if it was a members' show? Not the same birds here as there were there! No, they're not the same; they were a jolly sight better than any here," and so on *ad lib*.

Do not however despair; by a little judicious diplomacy on your part you can extract a lot of information from these two persons. The disappointed exhibitor will first attract your notice; he will be close to the class that you are interested in, talking loudly with some "pal" about the merits of his bird, or, on the other hand, giving a disdainful prod with his stick at the winner, the expression of his face showing the utmost contempt for the judge's award, and an utter indifference to any suffering that the winner may experience from the thrusting of his stick into its ribs.

Now is your chance. A mild remark from you to the effect that "I don't quite understand the judging of this class," will bring you the retort from the disappointed one, "I should think you didn't; just look at this," etc. You are his dearest friend for the time being, and he will go carefully through the whole class with you; very possibly he is viewing the birds through spectacles that are somewhat clouded, unless his own birds are under observation; still, note what he says, and, when you can, slip away and quietly seek the judge. Possibly, not being well up in the breed, you may have been so convinced by the plausibility of your previous companion as to believe he is in the right, and that the judge is quite wrong. You have the points of contention at your fingers' end. Put them to the judge, and see what he replies. It may be he will agree with all that you advance, but at the same time show you some fatal defect in your erstwhile friend's bird, which you did not notice, and which he had no intention of pointing out to you. Anyhow, if you are not thoroughly convinced by either party, you will have heard both sides of the question, you will have learnt where to look for defects, and what points of the breed are most highly esteemed, and the information gained at this show will put you on a far sounder footing to further prosecute your inquiries at the next show.

There is another way in which a great deal of useful information connected with fancy poultry may be picked up. We refer to the local fanciers' society. Join it by all means, for you will probably find many there able and willing to assist you with advice. There is one type of man connected with these societies that we would warn you against; not that there is any harm in him, but simply because it is only a waste of time to cultivate his acquaintance. He is one of those that has little or nothing to say about his own stock, or rather about the management and feeding of his birds; even when asked the direct question as to how he feeds his birds, he turns it off with "Oh, in the usual way," or something to the same effect. He aspires to be thought a "dark horse"; a man with "something up his sleeve"; a man that isn't going to tell everyone all *he* knows; not he. Our advice is, leave him alone, he is not worth your trouble. We have met several of such in our time, and when we have got to the bottom of their knowledge and supposed secrets, they were never worth a second thought.

Finally, the surest, and perhaps also the quickest way of getting "well up" in the points of any particular breed, is to purchase a fairly good specimen. Obtain the club standard of the breed, and carefully overhaul your bird. Having done this, enter him for a few shows. Attend these shows, provided also with the standard, and then carefully compare your bird with others in the same class, noting the number of points given in the standard for each item where your bird excels others, and also where he is deficient to them. In this way you will frequently be enabled to fully endorse the judge's awards, and gain a great deal of the information that you are seeking.

Was it fancy? or did we hear someone say, "This may be good advice to anyone wishing to learn the points of an exhibition bird, but what has it to do with the breeding of them"? Well, simply this, that if you have not got in your mind's eye the true type of bird that you wish to produce you will simply be groping about in the dark until you have got that ideal bird carefully stored away somewhere in your head. When once you have what we might term "taken hold of the ideal," it is wonderful how easy it is to spot certain birds which, although by no means perfection in themselves, still possess some marked characteristic which by judicious mating with your own birds eventually produces you infinitely superior stock. Whereas had you not the ideal engraved on your mind, you would probably have overlooked this one pronounced characteristic of the bird, simply because otherwise he was not a good specimen, and so have lost the opportunity when offered of greatly improving the value of your strain.

Starting a Strain.

There are several methods open to the would-be breeder of making a start, but in each and every case there are difficulties to contend with —difficulties that more or less disappear after a few years of careful breeding. Before, however, going somewhat fully into this subject of starting a strain, we would impress upon anyone the great importance of keeping a stud book, in which the parentage of the birds mated is put down, and of either rearing the chickens from different pens in separate places, or by marking the chickens when young in such a manner that their pedigree can easily be ascertained by reference to the stud-book. The importance of this arrangement will become apparent later on when we come to the subject of in-breeding.

Purchasing Eggs.

Perhaps the commonest method of first starting a strain is to purchase a sitting or two of eggs. Such a plan has, however, several drawbacks. Mr. So-and-So is advertising eggs at 7s. 6d. and 10s. 6d. per dozen, and from his special pen—from which he is breeding all his own—at 21s. per dozen. You make up your mind to go in for a dozen of the latter, but, unfortunately, there is no difference in the look between the guinea eggs and the 7s. 6d. ones. You must absolutely trust to the honesty of the vendor, which in poultry dealing, as in other businesses, is not a quantity that can always be relied upon. But supposing that you are dealing with a perfectly honest and straightforward individual—for we are pleased to say that the poultry Fancy contains many such—is it reasonable to suppose out of your one dozen eggs you will produce some tip-top specimens? We have known of many cases where a dozen eggs have been purchased at a big price from parties who would never dream of in any way cheating a purchaser not produce a bird fit to win even in a ten shilling selling class. But there is really nothing surprising in this, and certainly no grounds for imputing—as many an amateur often does—dishonesty on the part of the seller. Very possibly the vendor will again, the very same season that you purchased eggs from him and obtained no good birds, win almost everything; but very likely he has only about half-a-dozen real good birds, which he has obtained not from simply a dozen eggs, but from incubating several hundreds. If it were possible, or, rather, we would say, if it were probable that out of a dozen purchased eggs, two or three or even one real good specimen would be secured, then, instead of a guinea a dozen being a stiff price for eggs, ten guineas would not be a high price.

Nevertheless, we have known of several cases that have occurred to ourselves and friends, where a sitting of eggs have been sold, and the purchaser has produced one or more birds that were capable of beating all those produced by the vendor. Still, the chances are greatly in favour of second or third rate birds being produced instead of tip-top specimens. Again, it must not be thought because a certain exhibitor generally is well to the front with a certain breed that he of necessity knows all about the mating, and although he may consider No. 1 is his best pen, No. 3 may notwithstanding produce better. It by no means follows that because this bird is a Palace winner he will produce better or even as good stock as this other one that was never good enough to appear in the show pen.

On the whole, therefore, we consider that in the long run—as a general rule—the purchasing of eggs from which to start a strain is about the most expensive way to go to work, and are, consequently, unable to recommend this method to our readers.

A Mated Pen.

The next system adopted by many is to go to some well-known exhibitor and get him to mate a pen up for you. The chief drawback to this arrangement lies in the fact—this also applies to the case of purchasing eggs—that you are in complete ignorance as to the relationship of the birds mated up for you, and so once again you have to rely solely on the honesty of the vendor to tell you truly, as also on his straightforwardness in not mating up birds bred from two different pens that he knows when bred together will not produce first class progeny. Almost all breeds have certain members more suitable for breeding cockerels than pullets, and *vice versâ*, so that if he mates up hens from a pullet breeding pen with a cock from a cockerel breeding run, the offspring are not likely to prove of much value. Nevertheless, we have no hesitation in saying that if immediate results are desired this system is the best to adopt, always provided that the vendor *breeds* good quality birds, and that the purchaser is prepared to pay a good price for good quality stock.

Forming a Strain.

The plan that would commend itself strongly to anyone who is not partial to ready-made goods, and who desires that what he can accomplish shall be accredited to his own exertions and perseverance, is that of forming a strain on his own account. This method is decidedly the most fascinating of any, and, if perseverance and good judgment are brought to bear, is likely to prove the most remunerative in the end. But time is required. It is quite possible to purchase a winning cock and hen at the Palace Show, mate them together, and produce excellent offspring; but it is also quite possible—and in some breeds very probable—that such mating would never produce youngsters fit to grace the show pen.

By the words "forming a strain" we mean the purchasing of a pen not all from the same yard, but by obtaining a cockerel from one yard related as little as possible to the pullets procured from another or other yards, and by judicious breeding from these birds produce a strain of fowls eventually that have the same characteristics more or less marked in each individual.

The first year's mating may produce nothing worth having from a show point of view, owing to the infusion of entirely foreign blood, but each succeeding season, provided careful mating is resorted to, will show a vast improvement over the last; the percentage of exhibition birds bred each year will rise also, and in the end the champion of the season is perhaps your reward. This, bear in mind, is not obtained because you have a longer purse than your neighbour, but is the result of patient perseverance on your part. Anyone with money can buy a winner, but money alone can never show you how to breed a winner, and to be the breeder of a winner is looked upon by all true fanciers as a far more honourable position than being the mere possessor of the same.

The system that we would advocate to produce the above results when forming a strain will be given immediately after the subject of in-breeding.

TECHNICAL TERMS.

As reference may from time to time be made to technical terms, it is desirable to give a few short explanations of the same.

A STRAIN.

A good deal of confusion exists as to the term "strain." A strain implies hereditary disposition. Perhaps the best method of explaining what is commonly called a strain in the poultry world, is by supposing that we have three breeders, A, B and C, and that C is of the opinion that by crossing A and B's strains together he can more prominently develop or decrease, as the case may be, a certain point or points possessed by the stock of A or B in their natural state. If, therefore, C purchases stock from A and also from B, and if by the infusion of the blood of these two yards he is able to produce birds after several years'—more or less—careful selection and mating, that are hereditarily disposed to reproduce the required point or points, and which do so produce these points in a more or less marked degree in the progeny, C would then have established a "strain."

If C simply purchased A's strain, and bred only from these, even if he produced better birds than A did, they would still be A's strain. Many strains frequently differ so from one another that A's strain can easily be distinguished from B's strain.

A BREED.

A breed is a natural division of a race of fowls that differ in certain specific points from any other race of poultry.

A PURE BREED.

Considerable discussion has from time to time taken place in the Fancy Press as to what is a pure breed. A few will assert that the number of pure breeds can be counted on the fingers of one's hands, and that the remaining breeds are simply crosses or mongrels. But the reasons that they advance for coming to this opinion are very vague, and their arguments inconclusive. If the test of a pure breed consists in no cross ever having been employed, and if it has, that we should discard them and stick solely to those breeds that have not been crossed at some time or another, the Fancy, as well as commercial poultry keepers, would be in a sorry plight. Minorcas, Andalusians, Leghorns, Brahmas, dark Dorkings, not to mention scores of others, including the Old English Game fowl of the present day, would all have to go by the board. "This is rank heresy," we can imagine some of the old school saying; "the Old English Game is a pure breed, and always has been." We can only say that a few years since one of the most successful birds in the Old English Game classes in the south of England was bred from an Old English Game cock and a mongrel hen on a farmyard! The progeny can at the present day still hold their own. But the matter of employing an occasional

G*

cross to improve the breed from an exhibitor's point of view, is so general and so commonly employed by up-to-date fanciers (as anyone who has had much to do with the breeding of many different exhibition breeds at the same time would readily admit) that it is superfluous to further pursue the subject.

Besides, it is recognised by most naturalists that the common ancestor of all our poultry is *Gallus Bankiva*, and, therefore, as it has merely been by the selection of "sports," or through climatic influences, that we have any different breeds, and as even our oldest breeds throw a number of wasters viewed as show birds, we are of opinion that for all practical purposes, as far as an exhibitor is concerned, a breed that will produce a fair proportion of chickens of the desired type may safely be called a pure breed, whether made in England or Germany.

Below is given an illustration of the different points of a bird, together with an explanatory reference for the guidance of the amateur.

1. Comb (single).	7. Back.	13. Wing bow.	19. Hocks.
2. Face.	8. Saddle.	14. Wing bar.	20. Legs or
3. Wattle.	9. Saddlehackle.	15. Wing bay.	shanks.
4. Ear lobe.	10. Sickles.	16. Wing butt.	21. Spurs.
5. Hackle.	11. Tail coverts.	17. Flights.	22. Feet.
6. Breast.	12. True or hen tail.	18. Thighs.	

In the case of a Cochin or any similarly formed bird No. 8 would be termed the cushion.

VARIETIES.

A breed often has one or more "varieties." These, although differing somewhat in certain points to the original stock, still possess some of the characteristics of the breed. At times a variety is established by the selection and careful mating of some "sport," as so amply illustrated in the Sebright Bantam. But as a rule varieties are generally formed by judicious crossing and subsequent breeding.

PREPOTENCY.

Prepotency is the power possessed by an individual fowl of imprinting his or her likeness upon the offspring to the exclusion of the likeness of the other parent; which power is more apparent in some breeds than in others, as also is there considerable difference in the power possessed by individual members of the same breed. A strain of fowls that have been carefully bred for a number of years will be prepotent to a far greater degree than chance-bred fowls of the same breed. It is here, therefore, where the value of old-established breeds becomes apparent, inasmuch as they will, when crossed with another breed, stamp their likeness and qualities on the offspring to a far greater extent than a new breed would. But this is outside the question of breeding for exhibition.

MATING FOR SIZE AND SHAPE.

The male bird undoubtedly exercises a certain amount of influence in regard to the size and shape of the offspring. But to attempt to remedy—as so many amateurs do—the deficiency of size in their stock by the purchase of an extra large cock is the wrong way to go to work. The hen has far more influence over both the size and shape of the progeny than the male has. Take a broad-shouldered, deep-breasted cock and mate with narrow-shouldered hens, deficient also in breast, and the result of such a union will be but little, if any, improvement. Had, however, the tables been turned, and the hens possessed the size instead of the cock, far greater improvement would appear in the offspring. But, as we before remarked, the male bird does exercise a certain influence. It will be found that by breeding from large hens and a cock deficient in this respect the pullets produced show a far greater improvement than is observable in the cockerels, and it is only by continuing the process of breeding from large hens that the cockerels will far outdistance the original cock. There is no question but what the best plan is to have size and shape on both sides, but if a deficiency must occur on one side or the other do not let it be on that of the hens.

MATING FOR COLOUR.

Here we simply intend treating on whole coloured birds, and not with laced, barred, etc., in which we believe considerable influence is exercised by both sexes. It is useless to expect to breed good coloured youngsters from a bad coloured cock. Mate a good coloured black Hamburgh with only moderately coloured pullets, and many of the offspring will be excellent in colour; but mate a poor coloured cock with good coloured hens, and both cockerels and pullets will be deficient. As another example, take a white Leghorn cock, more or less naturally straw coloured, and mate with a carefully-bred strain of pure white hens, not a cockerel will be produced of pure white; and if the pullets are in-bred to the father, even the pullets in the next generation will show the straw coloured feathers. It is, therefore, of the utmost importance to see that the male bird is of such colour as is required to produce suitable colour in the progeny. We say advisedly, as is required, because, as will be shown presently, it may not be always desirable—with the exception of black or white birds—to breed from a standard coloured cock to procure the desired effect in the offspring. But what we would impress upon the reader is to take the greatest precaution to select his stock cocks with the view more particularly for obtaining the required colour in the youngsters.

The cock bird has far greater influence over the comb of the progeny than the hen. To breed from a male bird that is defective in comb is, in the majority of cases, to court utter failure, whereas to mate a cock good in this point to hens that are defective in comb will frequently produce good results. (See "Mating Minorcas," etc.)

Ear lobes, in our opinion, are about equally affected by the hen and the cock. That is to say, the proportion of increase in lobe of the progeny if a large-lobed male is mated to a small-lobed hen is about equal to the effects produced if a small-lobed cock is mated to a large-lobed hen.

Foot and leg feathering, and we would also mention the cushion in a Cochin, Orpington, etc., depends to a greater extent for its production on the male than on the female.

The eye, too, is decidedly more dependent for its transmission to the cock than the hen, and we would never advise a bad-eyed cock being employed for breeding, for personally we have always found them failures, whereas frequently have we produced every chick with good eyes from a bad-eyed hen bird. In stating this, it must not be supposed that we attach little or no importance to the hen's eye, as such is not the case; we would advocate the selection of good eyes on both sides; but although in the case of a cock having a bad eye we would, in the majority of cases, discard him from the breeding pen, yet, if a hen was similarly affected, but otherwise good, we would probably employ her for breeding purposes.

CARRIAGE OF TAIL.

Here the general tendency is for the cockerels to take after the father and the pullets after the hen. We mention this because in the case of our having an exceptionally good pullet-breeding cock, but with a poor carriage of tail, we would—unless closely in-bred to the hens—have no hesitation in mating them up for pullet breeding, provided that his hens excelled in this respect—the same law applying with hens deficient in tail carriage when mated to a cock exceptionally good in this respect for cockerel breeding.

SHAPE OF HEAD.

Again, cockerels are more liable to take after the father and pullets after the hen in the shape of the head, but our experience teaches us that the hen has considerably the greater influence of the two. But this, like many another point, greatly depends upon the prepotency of the birds. If the bird comes from a good-headed strain, and has been in-bred, more or less, for several generations, the power to imprint his or her likeness on the offspring will be materially increased whereas if he is merely a chance good-headed bird from amongst a number, and is mated with hens deficient in head properties, little, if any, improvement will appear in the progeny; it would only materially be shown by in-breeding the best-headed pullets with the father.

We would specially draw attention to the formation of the head. By this, more than any other given point, can you tell in the majority of breeds whether a bird has been carefully or loosely bred. Take, for instance, a Langshan ; we never saw a good Langshan Club type of head on a bird that had not many other points of the Club type to recommend her ; she may from some cause or other be under-sized, either from late hatching, sickness, or too close in-breeding, but still she always possesses other desirable Club type points. Again, take a first-class headed Game fowl; he may be weak in his legs and in-kneed, but still quality can always be seen. And so we might go on with most other breeds. The reason why the head, to our mind, is such an excellent criterion of good breeding is because both parents affect its formation to a considerable extent. The hen, as we have previously mentioned, has considerably the most influence over the shape ; the cock over the comb and eye. So that, in order to get an ideal head, both parents must possess some of the good qualities required, or else one of the parents of the bird under discussion must have been so carefully bred for this point for so long a time as to enormously increase its power of prepotency. In any case the bird would have good blood in it.

The Mating of Various Breeds.

Presuming that our advice has been followed by the tyro of making himself thoroughly acquainted with the points requisite in a good show specimen of the breed selected, the next point which he will be called upon to seriously consider is, How can I mate birds to produce these necessary qualifications? For, bearing in mind what we previously stated, viz., that the purchasing and mating together of the best birds of the season is not by any means a sure method of producing first-class progeny, the difficulties of successful breeding will at once be apparent to the novice. We have no hesitation in at once saying that in order to produce fixity of type in-breeding must be resorted to; but that alone will not produce good chickens unless the stock birds are judiciously mated; and although every breed differs somewhat in formation or colour from another, as also do sub-varieties of any breed, still, the main rules to be observed are often as equally applicable to other breeds as the actual one chosen for discussion. In order, therefore, to simplify the matter, it is our intention to, as far as possible, under the various headings, let our remarks apply equally to all breeds, unless in the course of this section special reference is made to some particular breed.

Mating Buff Breeds.

Unfortunately, at the time of penning these lines the shade of buff most desirable in a show specimen has not been definitely settled upon. But we think we are correct in stating that most, if not all, of our most prominent breeders of buff fowls are unanimously of opinion that the most desirable point to obtain is evenness of colour. No matter what shade of colour the buff may be, whether deep, light, or medium, the great aim that the breeder has to strive for is to get in whatever shade is chosen an evenness of that colour all over the body.

Personally, we prefer what we may term a "rich" buff; not a red by any means, but a sound, rich *buff*. We prefer this colour to the light or lemon buff because it is impossible—unless *red* buff blood is introduced from time to time—to maintain this shade of colour; the plumage will lose colour and a number of white feathers will appear, creating what is known as "mealy" plumage— in other words, a waster—either for the show pen or for breeding purposes.

We also object to the red-buff. In the first place because it is not a buff but a red, and secondly because we have never seen a red-buff cock even in colour; the wing-bow, and, more frequently than not, also the hackle, back, and saddle hackle are of distinctly a deeper colour than the breast and fluff. Notwithstanding this, such birds are often in the money, and, unless buff fanciers insist that buff and not red is to be their ideal, will probably continue to do so, for the simple reason that at the first glance they attract the eye and look more "showy." But when once the beauty of an *even*-coloured buff is properly understood it will always be ten times more appreciated than the red. It will thus be seen that our predilections are strongly antagonistic to, anyhow, the red-buff cock. Nevertheless, we consider him a useful member of society as far as the breeding pen is concerned.

At the majority of shows you will find that judges—at the present time, anyhow—"go for" a *rich* buff hen; again we would say, not one "patchy" with red on the wing, but a deep, even-coloured buff, and the deeper it is in colour the more favour it seems to find. Such a bird is not bred, or if it is it is only one such out of a quantity hatched—from an even-coloured buff, but from a red-buff cock, and that although a number of good pullets have thus been produced, the cockerels from the same pen were all more or less "patchy" in colour.

The system that we would advocate for breeding a buff, as distinguished from a red-buff cockerel, is this:—Choose a sound-coloured lemon cock—*i.e.*, a bird free from mealiness, even in colour, but with no approach to redness, and mate such a bird with good, sound, medium-coloured hens. By this mating the cockerels will be free from redness, and of a uniformity of colour quite unattainable when a red-buff cock is employed. The pullets, however, from such a union will probably be very "washy," if not actually "mealy."

It might with seeming justice be argued that if the above method is correct the same results would be obtained by mating a red-buff cock with very light lemon hens. But in practice it would be found that such was not the case as a general rule, the cockerels being far more influenced by the colour of the sire than they are by the colour of the hens.

We can well understand someone saying, "The best cockerel I ever bred came from a red-buff cock," but it must be remembered that there are apparent exceptions to every rule. We say apparent, because in the majority of cases there is a real cause why this seeming exception occurred. It might be that this red-buff cock was the progeny of medium-coloured birds on both sides, but a reversion to his former ancestors—as not infrequently happens—took place in his case, and that although in his individual self he was a red-buff, yet his breeding and his natural tendency would be to throw medium-coloured offspring.

Anomalies like the above happen far more frequently in large breeders' yards than they do in the case of a small breeder. The former, in the majority of cases, breeds, what we might term, loosely; he places different shades of coloured hens in each pen. It is very common to hear large breeders say, "Oh, I always like to mate up different shades together, so that if one doesn't hit it off another will." Whereas the small exhibitor has only room for one, or perhaps two, breeding pens, and consequently, as a rule, takes more pains to make sure that each bird placed in his pens is suitable for the production that he has in view. Only breeding a few birds each season, he probably has at his finger's end the pedigree of each as far back as their great-great-grandfather, and being in the possession of this knowledge, he is in a ten times better position to mate his birds successfully than the large breeder, who does not keep a stud-book and mark his chickens.

In mating for pullets we would choose a rich red-buff cock, as even in colour as you can find one, and mate with sound even-coloured rich buff hens. The cockerels from this pen will be very red in colour, and one and all more or less patchy, but the pullets will be exceptionally even and of good exhibition colour.

The above method of producing good-coloured birds is in contradiction to most of the large buff breeders that we have been in conversation with on this subject, and therefore we think, in justice to the reader, it is only fair that we should state that many prominent breeders are of opinion that the best cockerels can be bred by employing dark, or in other words a red buff, cock. But before passing judgment on the system we here advocate, we would strongly advise that at least it should be given a trial, and if a proper trial is given by anyone we are convinced that henceforth he will be an advocate of the principles we here lay down for the mating of buff breeds.

In concluding the subject of buff breeds, we ought to add that by the mating of even-coloured, medium birds on both sides the pen will produce a proportion of good birds of both sexes; but that the average result is not so satisfactory as when one pen is specially mated for cockerel and the other for pullet breeding. Further, that we like to see the buff colour extend right down to the root of the feather, not simply buff at the end of the feather, and when the feather is pulled back to find the lower portion white. Such birds are very liable to throw mealy offspring.

Mating Barred, Pencilled, Laced, Spangled, etc., Breeds.

Our contention is that the male bird has by far the greater influence over the ground colour of the feather, and the hen over the barring, lacing, etc. We do not wish to imply that the hen has no influence on the ground colour, and the cock none on the lacing, etc., because undoubtedly they have; but although cockerels are more liable to take after the father and pullets after the hen, we are convinced that the preponderance of influence over the ground colour of the feather lies with the cock, and the barring, etc., with the hen.

Mating Plymouth Rocks, etc.

As an example of the above contention, let us take the case of a Plymouth Rock. The chief defects observable in many cockerels is either that they are far too light in colour for exhibition, or that they are sooty in hackle, too dark on the wing bow, back, and saddle. This latter, to our mind, is caused by breeding from a dark cloudy top coloured exhibition cock and heavily barred exhibition hens, or birds approaching to this description.

The method we advocate for breeding Plymouth Rock cockerels is this:—Choose a good top coloured bird, i.e. a bird with good clear exhibition hackles, back, saddle, and tail, and free from any approach to brassiness on the wings.

We would not object a bit if he were too heavily barred on the breast, but would desire him to be of good exhibition colour on thighs and hocks, and withal decidedly what would be termed a good dark exhibition bird. Mate such a one as this with well-defined but narrow-barred hens. The barring should run right down to the root of the feather. This class of bird is only occasionally seen in the show pen, being, as a rule, a bit too light in colour (i.e., too fine in barring) to suit most judges.

You will observe, then, that we are discarding a cock bird that is in any way inclined to be sooty in the ground colour of his feathers, for the reason before stated that he has the greatest influence in this respect on the offspring; and that we are choosing hens that have a number of narrow bars on each feather, which number of bars will be reproduced to a very large extent in the progeny; thus you will obtain cockerels with a clear ground colour, but with a quantity of well-defined narrow bars on his top colour; and the heavy barring on the stock cock's breast will be rectified by the narrow barring of the hens. Pullets from such a pen will also be good, but hardly suitable to the present-day type.

As to the breeding of Plymouth Rock pullets, we would go to work in a totally different way. But before describing our method of procedure, we think it advisable to make a few remarks as to the class of bird it is required to produce. It is for this very reason that we have previously so strongly recommended a would-be breeder—or even an old-established breeder from time to time—to study well the winners at different shows. The fashion in the poultry Fancy is almost as variable a quantity as ladies' dress, and what a writer on poultry breeding may state to-day as the best means of breeding show birds may be entirely wrong to-morrow; that is to say, if he give a definite plan for the mating of each breed. Hence our reason for endeavouring to show, not simply the best means for mating different breeds for producing show specimens of the present-day type, but, if possible, to give reasons why such mating produces such results, in order that when fashion changes the same laws can still be applied, though in a different way.

Now, in Rock hens the majority of winners at the present day are broad barred-birds, with light ground colour. The barring is sharp and well defined, and of such depth of colour as to make a great contrast to the ground colour of the feather.

To breed such, we would take a very even-coloured cock, a bit too light for exhibition—not a washed-out waster, but a sort of bird that you would say, "That's a nice even-marked bird, but a wee bit too light for exhibition." Get him an even light shade all over, and free from any white in tail. Mate such a bird with exhibition hens, such as previously mentioned, or preferably with similar hens too heavily (but distinctly) barred for exhibition.

Here, again, the light ground colour of the cock will reappear in the offspring, and the evenness of his ground colour will be evident in the youngsters; but the broad and deep barring of the hens will not be lost —merely modified—and an even well-barred exhibition pullet will be the result.

Whether the above method is in accord with other breeders of Rocks or not we do not know; but, personally, we have adopted it with the utmost success.

MATING ANCONAS.

To give another example, showing the influence borne by the cock and the hen respectively on the colouring of the feather, we will take the Ancona, especially as the present type, or rather we would say colour of the bird seems widely different from that described in Mr. Lewis Wright's well known work, and because, owing to a loss of colour, there appears an exception somewhat to the general rule.

We have not the slightest hesitation in saying that a number of birds sold as Anconas originate from a cross between the black Minorca and white Leghorn; but between these and imported birds there is often a wide difference of appearance, though their ancestors were probably the same. Take the pullet as an example. The feathers should be black, with a kind of crescent in white at the tip. But the effect of a first cross between a Minorca and white Leghorn is to throw—amongst other colours—birds with pure black and pure white feathers, and many Anconas of the present day are feathered to a large extent with such feathers.

It must be remembered that with Anconas the ground colour of the feather is black, and that what variation there is in the feather is simply a *loss* of ground colour.

This is a totally different case to a white or golden ground colour and a black lacing or spangling. If the Ancona is deficient in the ground colour (black), then the progeny will be disposed to take after him more than they will after the hens. Especially is this the case when colour has degenerated to white. But as we have several times repeated, cockerels are more liable to take after the male bird and pullets after the hen, and the mating that we advocate is as follows.

For pullet breeding, choose a cock of even colour, but a bit too light for exhibition, and mate him with hens too dark for show purposes.

For producing cockerels, choose again as even a colour bird as possible, rather too dark for exhibition, and mate with hens of exhibition standard. It must be borne in mind that when once colour is lost—as it is to a certain extent in an Ancona—the tendency is to go on in the same direction, and lose more colour, therefore, although an exhibition Ancona cock mated with show hens may, and probably will, produce some good show cockerels, still the tendency is to throw them too light.

We would here remark that we think it most desirable that Ancona breeders should endeavour to perpetuate the beautiful green gloss found on some Anconas, as it undoubtedly adds considerably to the appearance of the bird.

Now, in the above mating we still carry out the principles advocated previously. In the case of breeding Ancona pullets, there is no question arising of the hen having greater power than the cock to imprint a bar, spangle, or lacing on the feather, it is simply a matter of the male having the greater control over the ground colour. And, as it is desirable that the pullets produced should lose ground colour at the end of the feather, and as they are liable to take after the hens more than the cock, we choose a mate somewhat deficient in ground colour, and mate to hens that have not lost enough.

Then with regard to the breeding of Ancona cockerels, the only apparent inconsistency to our system of mating lies in the fact that we advocate breeding from a cock that has too much ground colour, but this is done simply to counteract the natural tendency to lose colour in all breeds that have any white in their plumage.

Mating Andalusians, etc.

As another illustration of our system of mating laced, etc., breeds, we will take the Andalusian. For breeding cockerels we would choose a good exhibition cock—one possessing exceptionally good clear ground-colour on his breast, though the lacing might not be very distinct, and also having a good rich top-colour. Mate such a bird with hens possessing well-defined, sharp lacing on the breast, the same running well into the fluff. If the hens are too dark on the wing bow, hackle, and saddle, so much the better.

For pullet breeding choose a cock too light in top-colour, but of a good blue ground-colour—the counterpart of the ground-colour that you would desire to see your pullets become. Mate this bird with exceptionally well-laced hens—not necessarily exhibition specimens (they may be too dark in ground-colour for this purpose), but when the feathers are closely examined the lacing should be sharp and well defined.

Again, it will be seen that we apply the same law as in the other cases mentioned. Still, it is only right to add that the more perfect the specimens are in those points that you wish to produce in the offspring the better will be the result. As, for instance, in regard to the Andalusian cock for cockerel breeding, we would prefer that not only should he possess a good clear ground-colour to his breast, but also that the same should be well laced; as also, in the matter of pullet breeding, we should prefer to select hens, if we could obtain them, not only with exceptionally good dark lacing, but also of a good, rich, clear ground-colour. Nevertheless, as we said when commencing this subject, we are firmly of opinion that the male bird has by far the greater influence over the ground-colour and the hen over the lacing, barring, etc., so that, although it is frequently desirable to obtain in both parents the points desired in the offspring, such as a gold or silver Wyandotte cock for pullet breeding being laced like a hen on other parts than simply the breast and fluff (which example is a proof of the soundness of our system, because it means that the ground-colour of the feathers is the same as are desired to be reproduced), yet the main considerations that we would insist on in mating is that the male bird must possess such ground-colour in his feathers as is desired to be reproduced, and that the hens should be the possessors of such barring, lacing, etc., as may be required in the offspring.

MATING PARTRIDGE COCHINS.

As a final illustration of our system, and as a further proof of its correctness, we take the case of the partridge Cochin. It has for many years been looked upon as a waste of time to attempt to breed good exhibition partridge Cochin cockerels from a good exhibition partridge Cochin hen. But on the contrary, you choose hens almost devoid of pencilling, and mate with as rich red coloured cock as you can find. The male has the requisite qualifications for producing the required ground colour, and the hens are minus the lacing, or anyhow possess but little.

For pullet breeding, on the other hand, you select as well pencilled hens as you can possibly find, especially having regard to solid and distinct lacing, and mate with an orange red cock, splashed on the breast and fluff with reddy brown feathers, as it shows a tendency to revert to the ground colour desirable in the pullets, and the richer the cock bird is in top colour, provided he is more or less splashed on the breast, the deeper and richer will the ground colour of the pullets be.

Here again, then, the cock possesses the ground colour, or decided tendency to the ground colour, required in the pullets, and the hens have the pencilling.

MATING MINORCAS, LEGHORNS, ETC.

We have previously treated on the subjects of obtaining shape, size, lobe, colour, etc., and the principal reason why we give a paragraph to Minorcas, etc., is that in breeding large-combed birds further consideration in regard to the stock birds' combs is necessary beyond the points we have already mentioned.

About the best pullet-breeding white Leghorn cock that we ever possessed we purchased for five shillings ! simply because his comb had fallen completely over and his owner considered him in consequence of no use. But if your Minorca, Leghorn, Andalusian, Ancona, etc., pullets are deficient in comb, then try the mating of them to a cock that has what is known as a large "beefy" comb, so heavy that it has not sufficient strength in it to stand upright, but falls completely over, and undoubtedly you will observe a marked improvement in the pullets produced from this mating. We would, however, strongly advise that such a bird should be dubbed. Repeatedly have we seen birds with such heavy combs that it was with great difficulty the bird could hold its head in an upright position, and from the great weight of the comb inconveniencing and causing such "splitting headaches" that the fowl was literally pining away; sullen, discontented, no life in his movements. Dubbed, and within three or four days the picture of health and happiness.

On the other hand, to attempt to breed cockerels from a naturally weak-combed cock is folly. Choose a bird with a small, or anyhow only a medium comb for this purpose, and if the hens' combs are inclined to stand up a bit so much the better.

The cock's comb should be broad at the base and firmly set, and instead of coming away at the back of the head, which some do, and which is not of so much importance for pullet breeding, it should nicely follow the curve of the head, but at the same time not be so close as to actually touch the back of the neck.

As regards the general cause of "comb over," "thumb marks," and the method of promoting the growth of combs and lobes, these matters we have dealt with in Section VI., "Preparing Fowls for Exhibition." We mention this particularly because it will be observed in the preceding paragraph we have been careful to say "a naturally weak-combed cock," for many a male bird that possesses a weak comb does so not because its breeding had a tendency to throw weak combs, but simply through mismanagement as a youngster, either whilst with the hen, in the incubator, or when removed to the chicken house.

In mating brown Leghorns choose for cockerel breeding a bright coloured exhibition bird—not what would be termed a dark bird, but one of a good bright colour—and mate with light or medium coloured hens that have an inclination to lose the pencilling by being a bit red or "foxey" on the wing.

For pullet breeding choose a cock more even in colour than an exhibition specimen, his hackle, saddle and wing-bow being very similar in shade, and of a deeper, duller colour than the exhibition bird, and mate such a one to good exhibition hens, light in colour, distinctly pencilled, free from rust on wing—in fact a good partridge colour right through.

IN-BREEDING.

We now come to the very important question of in-breeding. "Be sure that the cockerel is not in any way related to the hens that you send me." Such has been the sum and substance of hundreds of letters we have received in our time. Even at the present day an enormous amount of prejudice is shown by breeders, not only of poultry, but any other kind of live stock, when the subject of in-breeding is touched upon.

Before, therefore, going into the pros and cons of in-breeding, we would like to give one or two examples of the results of in-breeding that have—amongst many others—come under our own personal observation.

A well-known sheep-breeder in the South of England, whose flock often numbered six thousand head, every autumn used to hold a sale of ram tags (i.e., rams rising two-year-old), and invariably obtained good prices for the same. The method that he adopted was to each year buy one or two of the best rams that he could find of a foreign blood to his own flock; these he mated with the pick of his own ewes, but owing to the large number of ewes that he possessed, he was more or less compelled to use some of his own blooded rams, and these were placed with his second-rate ewes. Now he was strongly of opinion that the best ram lambs that he produced were procured by his picked ewes and his purchased foreign blooded rams, but we maintain that the only reason why he obtained good typical stock was from the fact that he used a number of his own bred rams with his second rate ewes, and that had he placed his picked ewes with his own blooded rams the result would have been infinitely better, as the sequel to this story will show.

At his death his eldest son took over the management, and owing to certain pecuniary reasons the flock had to be reduced to about as many hundreds as formerly it was thousands. The son had been trained all his life to the business, and naturally reserved the best of the ewes for his own flock, and being fully acquainted with his father's system of breeding, and having now such a much smaller stud, he decided to each year to buy sufficient of the best foreign blooded rams for running with the whole of the flock.

Five years did he thus continue to breed, and each year did the quality of his flock decrease. Not only was type to a large extent lost, but the staple of the fleece had deteriorated; and taken as a flock they looked what we should term in the poultry fancy a mongrel lot.

At this time we had a long talk over the subject of in-breeding, and, after considerable persuasion, we succeeded in getting him to give the matter a trial. The first year's result showed considerable improvement, the second more so, and the third year another well-known breeder of the same variety told him that he never remembered seeing a better lot of ram lambs even in his (the son's) father's time.

Another of our earliest recollections of in-breeding occurred with Flying Homer pigeons. Our brother bought a pair of silver Flying Homers, which were kept at a small outlying farm right away from any other loft, and, to our certain knowledge, no fresh blood was ever introduced either intentionally or accidentally. In this way they in-bred for six years. We removed after this period a couple of young squabs, and with one of them won several matches, including a 150 miles race, thus clearly proving to our mind that not only is fixity of type maintained, but that even strength and stamina are not of necessity ruined by the process of in-breeding.

The effect of in-breeding is not only to establish a high degree of prepotency, but also to enlarge in the offspring certain peculiarities of form or latent tendencies possessed by the parent stock.

As, for instance, supposing we take two birds possessed of very large white lobes—say, brother and sister—and mate these two together, we shall probably produce a number of chickens that will be possessed of considerably larger lobes than either of the parent stock; and if we continue in like manner to in-breed, not only the lobe but the whole face would eventually become white.

But exactly similar results would happen if the parent stock had some tendency to disease. The first time of in-breeding would produce progeny more liable to this disease, and perhaps in the next generation the disease would develop and destroy a large number. But such consequences are not obtained simply from the fact of in-breeding, but from in-breeding from unsuitable stock.

Certainly, if we desired to put stamina into our stock we would not think of resorting to in-breeding; but we have no hesitation in saying that many a yard at the present time would possess far healthier birds if in-breeding had been resorted to instead of their owners having year after year imported fresh blood. Poultry-breeders are not nearly particular enough to ascertain the health of the stock from whence their imported stock cock is derived. It may be that this healthy-looking bird is bred from stock infected with diphtheritic roup or tuberculosis—that, in fact, he was one of the few of the brood that managed to struggle through and survive, and that although no signs of the malady is apparent, his system is more or less infested with the germs of the disease, and which disease will be transmitted to the youngsters, many of which will not be strong enough to withstand its ravages. Undoubtedly, under such circumstances, in-breeding from one's own stock that is known to be strong and hardy would have produced infinitely better results than the importation of fresh or foreign blood.

Decidedly the productiveness of the bird is not increased by in-breeding; but even here the theory that it is wrong to in-breed is carried to much greater lengths than it ought to be. We will suppose that you have a flock of, say, Houdans, which are first-class layers. Now, if you continue to in-breed these birds, egg production will be checked to a considerable extent. But a far quicker way of checking the supply of eggs—though not to the extent that it could be done by a continuance of in-breeding—would be by importing into your yard a cock derived from a very poor strain of layers. The effect would be immediately apparent in the offspring.

Even for the production of eggs, if we had a first-class laying strain we should prefer, to a certain extent, to in-breed in preference to obtaining a fresh-blooded cock that had nothing outside his fresh blood to recommend him as a layer.

But when we come to fixity of type, as in a show bird, where the comb has to be possessed of certain characteristics, where the shape of the head is fixed, the colour of eye, length of body, carriage, colour of plumage, legs, beak, and markings all defined in a recognised standard, then we say it is an impossibility to breed a fair proportion of such birds in a yard where in-breeding is not resorted to.

When to In-Breed.

Because we know from experience what hard work it is to breed good birds and animals with a fixity of type unless in-breeding is resorted to, let not the novice suppose that we advocate the system under all and every circumstance. Supposing you possess a second-rate stock of birds, it would be very stupid to mate up your hens with one of the cockerels produced by them; it would be far preferable to purchase a real good foreign blooded bird and put with them. Or even supposing that, taken all round, your stock is good, but at the same time possesses some fault—even though comparatively slight—common to all, it would be folly to in-breed, because the fault would be magnified in the offspring; and even if the fault was not common to all, it would be unwise to in-breed any members that did possess such fault, or to in-breed such stock at all if the fault was apparent amongst a number of the stock. So also would it be folly to in-breed from any specimens not in robust health.

The only cases in which we recommend in-breeding are these:—Firstly, for the purpose of fixing some chance good point produced; secondly, for the improvement of good birds; and, thirdly, to maintain the high state of perfection that certain fowls may have been brought to.

With reference to the first of these cases, the fixing of some chance good point, the following is written by a well-known and much respected specialist of a certain breed. "Above all things, always breed from the best birds, and although a bird with only one particularly good point may take your fancy, avoid it for breeding purposes—a fairly good all-round bird will produce better results than one that is perfect in one point, and faulty in all others." The latter portion of this quotation, viz., that a good all-round bird will produce better results than one only perfect in one point, is correct. But to advise the breeder never to breed from a bird that is perfect in one point, simply because it is faulty in all others, is cutting nine-tenths of the ground from under the feet of the one who desires to improve any given breed.

We could quote a number of instances where we personally have bred from a bird simply because he had some point in perfection, though more or less faulty everywhere else, with the utmost success; but perhaps the following is the most striking example because the bird chosen was an utter waster (from a show point of view), and a cross-bred fowl into the bargain. The day after our arrival at a certain poultry farm to take over our duties as manager, we came across a poultryman killing some fowls for table purposes, and, whilst chatting with him, he drew out of a sack by his side a cross-bred pile Leghorn and Brahma, and it was only by quickly calling out to him we were enabled to prevent the bird's neck being broken. It was a "stitch in time" that saved a good many more than nine—winners. We further learnt that orders had previously been given that this bird should be killed several times, as its owner disliked seeing such a mongrel object running about, and that it was entirely owing to the neglect of the poultryman in not performing his duties properly that the bird had not been put out of the way weeks before.

A brief description of the bird was as follows:— Comb, small and mean; head, Gamey; body, long, with abundance of fluff at the stern; legs, short (or apparently short, on account of fluff and feathering); hocks, vultured, stiff feathers standing out some four to five inches; shanks, well feathered, the feathering running well down the outer toe, and a few on the middle toe breast, very "marbled." As in the quotation previously given, there was only *one* good point, and that was colour; all the rest were not only faulty but downright bad.

The first year of mating this crossbred bird to true-bred pile Leghorn hens produced a cockerel good enough to obtain reserve at the Palace Show. The following season this cockerel was in-bred to the parent stock, the progeny from which mating swept the decks at the Dairy, Palace, Birmingham, and Club Shows, besides winning prizes at many minor exhibitions.

As we said before, this is the most striking incident of success from choosing a bird possessing but one point in perfection and the rest faulty, that we are able to give from personal experience; but we have to a less degree experimented many times with numbers of other breeds, and almost always with the utmost possible success, though the process may at times have to be carried on for a much longer period than in the case mentioned.

In the second case when we would recommend in-breeding, viz., for the purpose of improving good birds, we will take as a sample one of many such cases as have come under our observation—the amateur exhibitor who doesn't mind spending a bit yearly over his fowls but desires to win a few prizes. He has always a decent lot of birds to look at, but yet, as the yokel says, "he don't seem to get no forrarder." In many, many such cases we are convinced that the sole reason is that each season he sees a cock bird that he likes, thinks it will just about hit it off with his hens, and so purchases foreign blood year after year. Were he to select the best young cockerel that he had reared, and mate to his stock hens, and place his stock cocks with the best of the pullets produced, over and over again would he be miles nearer the goal that he is aiming for than he would be by the purchasing of foreign blood.

Lastly, where we recommend in-breeding is in order to maintain the high state of perfection to which the stock has attained. Many a yard has taken years to regain the position that it lost, solely because its owner was tempted by the good looks of a certain cock to introduce entire foreign blood. At the present day the evil results likely to accrue by the introduction of a male from another yard are not nearly so great as they were formerly. Railways, innumerable poultry shows, and general facilities offered for the intercourse of fanciers and for the interchanging of stock are now so great that thousands of yards have stock more or less related one to the other. But supposing you have, say, a first-class Andalusian hen, and you mate her to an A1 Andalusian cock of entirely foreign blood, though to all outward appearance no better bird could be found for mating with your hen, you would most likely find that not only were the majority of the chickens produced either black or white, but those that did come blue were not to be compared either to the father or the mother, the entire fresh blood causing, as the great naturalist Darwin says, a reversion to their former ancestors. It is the same with any breed. The strain as a strain is useless; it has lost its characteristics; and certain bad features, which by a long course of careful selection and in-breeding had disappeared from the strain, crop up in all directions: foul feathers, faulty combs, bad coloured legs, wrongly shaped heads.

IN-BREEDING.—HOW TO FORM A STRAIN.

Were we personally about to form a strain we would desire to commence with not less than three pens of birds, the hens in each pen being of the same strain and the cocks procured from three separate strains entirely. For by this method and by carefully marking the chickens and keeping a strict account in the stud-book of your mating, in-breeding may be carried on for many years without the necessity of importing foreign blood. We will suppose, then, that a start is made on these lines. All the pullets are purchased of A's strain and divided into three separate pens; that a one-year-old cock is purchased from B (see remarks on "Suitable Ages to Mate Together"), another one-year-old cock from C, and a third from D The first year's mating and results would be thus :—

MATING.	RESULTS.
B cock mated to A hens will produce	One-half B and one-half A's blood.
C cock mated to A hens will produce	One-half C and one-half A's blood.
D cock mated to A hens will produce	One-half D and one-half A's blood.

SECOND YEAR'S WORK.

Retain a few spare BA, CA, and DA cockerels in case of accident, or for future breeding, and mate :—

BA cockerels to mothers (or hens in next pen).
CA   „   „   „   „   „
DA   „   „   „   „   „
BA pullets to B cock (father)."
CA   „   C   „
DA   „   D   „

SECOND YEAR'S RESULTS.

BA cockerels will produce	..	$\frac{3}{4}$ A and $\frac{1}{4}$ B blood
CA „ „ „	..	$\frac{3}{4}$ A and $\frac{1}{4}$ C „
DA „ „ „	..	$\frac{3}{4}$ A and $\frac{1}{4}$ D „
BA pullets „ „	..	$\frac{3}{4}$ B and $\frac{1}{4}$ A „
CA „ „ „	..	$\frac{3}{4}$ C and $\frac{1}{4}$ A „
DA „ „ „	..	$\frac{3}{4}$ D and $\frac{1}{4}$ A „

THIRD YEAR'S WORK.

It will now be observed that we have in-bred so that we have birds possessing three-quarters of the blood of either A, B, C or D, and having got so far, we are in an excellent position to determine which is the best "blend" to work on. Possibly we may find that the infusion of D and A's blood shows to the best advantage, and, consequently we decide to make the DA birds the foundation of our strain. But it may be thought that once more directly in-breeding to D cock will give still better results, and for this reason it is advisable to at first commence where possible (see "Suitable Ages to Mate Together") with a cockerel instead of a one-year-old cock. And, supposing this has been done, or that the cock has been properly managed during the off seasons (see "Number of Hens to each Male"), and in consequence is still more or less a vigorous bird we would mate him thus :—$\frac{3}{4}$ D and $\frac{1}{4}$ A pullets to D cock, will produce $\frac{7}{8}$ D and $\frac{1}{8}$ A.

Now, although we have at times in-bred still closer than this with beneficial results, it nevertheless stands to reason that by too close in-breeding anything but desirable results are likely to accrue, and therefore we would not in the general way advocate closer in-breeding than seven-eighths, especially as we consider for all practical purposes—unless in an exceptional case—this amount is amply sufficient.

But it may be our opinion that in the third year's mating direct in-breeding to D is not desirable, or the bird may be sterile, or even dead, and so the above quoted seven-eighth D and one-eighth A birds are never produced, but instead we mate up as follows :—

$\frac{3}{4}$ D and $\frac{1}{4}$ A cockerels to DA pullets (now one year old).
„   „   „   BA   „   „   „
„   „   „   CA   „   „   „
„   „   pullets to DA cocks   „   „
„   „   „   CA   „   „   „
„   „   „   BA   „   „   „

THIRD YEAR'S RESULTS.

$\frac{3}{4}$D and $\frac{1}{4}$A cockerels to DA hens will produce $\frac{5}{8}$D and $\frac{3}{8}$A's blood.
„   „   „   BA   „   „ $\frac{3}{8}$D, $\frac{1}{8}$B, and $\frac{4}{8}$A's blood.
„   „   „   CA   „   „ $\frac{3}{8}$D, $\frac{1}{8}$C, and $\frac{4}{8}$A's blood.
„   „   pullets to DA cocks   „ $\frac{5}{8}$D and $\frac{3}{8}$A's blood.
„   „   „   CA   „   „ $\frac{3}{8}$D, $\frac{1}{8}$C, and $\frac{4}{8}$A's blood.
„   „   „   BA   „   „ $\frac{3}{8}$D, $\frac{1}{8}$B, and $\frac{4}{8}$A's blood.

Fourth Year's Work.

We have now arrived at an epoch in our breeding, for, no matter how we mate, we cannot produce birds with more, in fact not so much, blood of D in them as the $\frac{2}{3}$D and $\frac{1}{4}$A birds previously produced in the second year. Let us mate the greatest amount of D's blood at command, viz., $\frac{3}{4}$D and $\frac{1}{4}$A cockerel (now one year old) to $\frac{3}{4}$D and $\frac{1}{4}$A pullets (now one year old). The result is exactly the same. If we mate $\frac{3}{4}$D and $\frac{1}{4}$A cocks to $\frac{5}{8}$D and $\frac{3}{8}$A pullets, we get a fraction less of D's blood than that possessed by the stock cocks; we have practically formed a new blood E. It may be considered advisable to mate up in this manner, or, on the other hand, it may be thought necessary to introduce fresh blood, in which case we have birds that not only merely possess $\frac{5}{8}$ of D's blood and $\frac{1}{8}$ of A's, but, also, have $\frac{2}{8}$ of entirely fresh blood, viz., B or C; and so by judicious mating we can carry on our stock without having to resort to outside blood for a considerable time.

Another Example.

Neither your accommodation nor your purse will perhaps admit of such extensive operations as the foregoing plan would involve, and so we give another example on a smaller scale. Purchase a one-year-old cock D, and three pullets A, B, and C, the pullets being of a different strain to D, and also to one another. Exactly the same rules can be carried out in this instance as in the preceding case. By carefully watching your hens for a little while you can soon tell which eggs are laid by A, B, or C, and by hatching A's eggs under one hen, and B's and C's under others, and then marking the chickens afterwards, their parentage can easily be known, and in this way can you form a strain—your own strain—which will possess certain well-defined characteristics, and be prepotent to produce similar results.

The Introduction of Fresh Blood.

All breeders know that whether the years be few or many, the time will assuredly arrive when the introduction of alien blood becomes an absolute necessity. If you place an entirely foreign blooded cock into your yard (in the case of your possessing but one pen) it probably means one lost season anyhow, if not more. Naturally, if you have a number of birds, it would be advisable to put up a pen of good birds to a foreign-blooded cock, that is, if you can spare them. But in a small yard this is impossible, and the best way to go to work is to obtain a good foreign-blooded hen. Keep her eggs separate from the others, and mark her chickens. Do not feel disappointed because her chickens are by no means up to the mark compared to your own strain; it is simply the reversion that has taken place, owing to the introduction of fresh blood, to, perhaps, long lost characters. And by choosing the most suitable of the youngsters to in-breed into your own strain next year, all will again come right.

Another common method of introducing fresh blood, and one of the best if you can rely on the vendor, is to purchase a bird from some fancier possessing good stock, and to whom you have previously sold the bird that bred the one you wish to buy. A bird possessing half the blood of your own strain and half that of some one's else is amply sufficient. But as to when fresh blood should be employed it is difficult to state, i.e., to give the length of time that in-breeding may be carried on, so much depends upon circumstances.

If the stock is losing size, if they do not appear to be healthy, or if they are developing some fault, by all means stop in-breeding, even though you have only resorted to it for one season. As we have previously remarked, the effects of in-breeding is to increase certain characteristics possessed by the parent stock; to enlarge the prepotency of the birds, and also to develop latent tendencies to disease should the stock birds be predisposed to such. But so long as no ill-effects are noticeable we would object to the introduction of fresh blood in a tip-top strain of exhibition fowls, so long as we could breed our chickens with not more than seven-eighths of the same blood in their veins, or until we were virtually breeding the same blood over and over again.

Taking it, however, that the time has arrived when it is thought advisable to introduce fresh blood into your strain, the chief point to consider is not so much the advisability of choosing a first-class exhibition bird, as it is to carefully note in what respect your strain fails somewhat. And, further, to bear in mind whether it is the hen or the cock bird that has the greatest influence over that particular point which is faulty in your birds, and which you desire to rectify. If your birds fail in size, then you will purchase one or more hens; if comb is the weak point, then you will obtain a male bird, and so on with any other point that your stock may fail in, naturally obtaining the best all round bird you can, but anyhow make certain that he or she excels in that particular point in which your own stock fails.

SUITABLE AGES TO MATE TOGETHER.

Too much stress is, we consider, often placed on the well-known advice, "Mate cockerels to hens, and pullets to one or more year-old cocks." In our opinion the advisability of adhering to this rule greatly depends upon circumstances. It depends upon the breed, and as to the forwardness of the birds at the time we desire to breed from them. A breed that quickly comes to maturity can with safety be bred from much sooner than a breed that is a long while before maturing. For instance, take a Brahma or Cochin: these often grow until they are eighteen months, and even older; but a Minorca or Leghorn will be "set" by the time he is ten to eleven months of age. Not only this, but different strains of any given breed vary considerably as to early or late maturity, and, of course, the feeding, rearing, and general management of the youngsters make a vast amount of difference. But, as a general rule, it may be taken that in the heavy breeds, such as Langshans, Brahmas, Cochin, Dorking, Indian Game, etc., the season's youngsters are best mated up with older stock. Nevertheless, we have repeatedly seen eminently satisfactory results even in these breeds, when the season's produce has been mated together. Early hatched, say January or February, and well-grown Dorkings will in the following March or April produce as strong, healthy, and quick-growing chickens as older birds will; and, naturally, this applies with still greater force to such quick-growing breeds as Leghorns and their like.

No doubt the reason that brought forth the condemnation of breeding young stock together, and the advice that they should always be paired with birds older than themselves, was that young birds were indiscriminately mated together. But if the young stock have been properly reared, have had no check from disease or insects, and are well-grown birds, we would just as soon breed from them as we would if they were mated to birds older than themselves. We mention this not because we would advocate the breeding of young stock together as the general rule to follow, but to show that if occasion arises when it is necessary to either breed in this way or not to breed at all, we should not have the slightest hesitation in so mating.

NUMBER OF HENS TO EACH MALE.

Different breeds require different numbers of hens to a given male bird. One might run double, treble, and even more hens with an Old English Game than would be possible with a Brahma.

The accommodation of the birds, too, makes a considerable difference. A bird that may safely be given a dozen wives whilst possessing complete liberty would not, probably, fertilise the eggs satisfactorily from more than half that number in an average confined run; and if the run was exceptionally small, even this number might prove too many for him.

A one-year-old bird that has been used for breeding purposes should not be given as many hens as a cockerel; and a two-year-old bird not so many as a one-year-old. A one-year-old bird that is removed from the hens when the breeding season is over, say the end of April, and placed in a cock-box as advocated in Section IV. ("Housing and Feeding of Stock") until the breeding season comes round, will probably prove as fertile as most young cockerels; and a two-year-old cock similarly treated would, in the majority of cases, fertilise as many, if not more, eggs than a one-year-old bird that had been left to run the whole summer with the hens.

The time of the year, and the kind of weather one is getting at that season—for we can never tell in advance—makes a great difference. In cold, rainy weather—which is the worst kind of weather for obtaining fertile eggs—the cock bird should be given considerably less hens than in cold, bright sunshiny weather, and half as many more hens may be allowed him (or even double) in bright warm weather.

Individual birds of the same breed and of the same age are so differently constituted that although a score of hens might not be too many for one, half a dozen would be too much for the other.

Notwithstanding all these "ifs" and "ands," we have been asked to state definitely how many hens should be allowed to each cock of the different breeds! We respectfully decline to do anything of the sort. In order to give sound advice on this matter we should have, with each breed, to go into all the details previously mentioned; and even then we should still be confronted with the fact that "individual birds of the same breed and the same age are differently constituted."

CUTTING OUT OF FEATHER.

In breeding Brahmas, Cochins, Pekin and white-booted Bantams, as also with Polish, Houdans, etc., one great cause of unfertility is owing to over-abundance of feathering. With Brahmas, Cochins, and Pekin Bantams, many more fertile eggs can be obtained by taking a pair of scissors and trimming the fluff close to the abdomen, both with the cocks and hens, and clipping off the foot-feathering. Naturally, you cannot do this with birds that you desire to show; but birds should not be shown when once they are placed in the breeding-pen, and when the season is over you have not many months to wait before the moult takes place.

As to Polish, etc., by trimming up the crests so that the birds can see properly, their nervous system does not receive those shocks that it is continually subjected to when such is not performed, and considerable benefit in fertility of eggs will thus be gained.

BREEDING BANTAMS.

It is impossible for us, before concluding the subject of this section, to omit saying a few words about Bantams. The would-be poultry fancier who has insufficient space for the keeping of large fowls has surely a few square feet on which to erect a Bantam house and run. Although small, little things are not always to be despised—sovereigns are not nearly so large as pennies. Hitherto we have not written for the fancier who merely keeps his birds as pets, and have no intention even now of dwelling on the prettiness of Bantams. We have always endeavoured to write for those—by a long way the great majority—who desire to make their poultry pay. That a good and popular strain of Bantams will pay is easily proved. Accommodation that they require is very little; amount of food that they eat is—to use a common expression—a mere song; the attention necessary is no more than with larger fowls, if as much; cost of carriage to and from a show, about a quarter of that of an ordinary fowl; cost of entry at a show and the prize money that they can win equal to their larger brethren; value of a good Bantam, about equal to a large fowl.

In breeding Bantams the chief aim is to keep the size small; therefore, you will pay more attention to obtaining small hens than a small male bird, though, naturally, you will get both as small as possible. In order to keep down the size there are two important matters to be considered, viz., when to hatch and what to feed on.

Many breeders hatch late in the season, so that the chicks do not mature till spring. But, although this does check their growth, it often means at the loss of other properties. A Black Rosecomb will probably have a small, stunted, mean tail; a Game Bantam be wanting in reach, and so on. You have probably observed that in large fowls—this depends a good deal on the season—often the January-hatched birds are, later in the year, dwarfed by those hatched in March and even April. Taken all round, we consider the best months for Bantam-hatching are February, the end of April, through May, and until the middle of June.

As to the feeding, don't overfeed, but give amply sufficient. Rice, boiled nice and dry, so that each grain is separate from another, in half milk and half water, is a capital feed for them. For hard food, when very small, millet, canary and a little broken groats are excellent; at about a fortnight to three weeks old, wheat should form the principal item of corn, with a little canary seed added, and when the groats are thus stopped, the boiled rice may have just a little oatmeal mixed with it as a strengthener.

Breeding from really good stock birds, there is no occasion to cruelly starve the youngsters, as some breeders do; by choosing the food so that but little bone-forming material is supplied, the reults are more satisfactory than a course of semi-starvation. Bread, first scalded with boiling-water, and then milk added to it, makes an occasional suitable change.

The foregoing method of feeding is suitable to any variety of Bantams; but with such birds as Game, Indian Game Bantams, and their like, we prefer, after about three weeks or a month, to give no soft food at all, or but just a little occasionally, as the giving of much soft food has a tendency to create looseness of feather. Wheat, with a little dari, is about the best staple food for growing stock, and half wheat and small white New Zealand oats, weighing about 46 lbs. the Imperial bushel, for matured birds that we desire to keep in exhibition trim.

Although we do not advise starving the birds, still, do not go to the other extreme. Not many months ago a certain Variety Bantam fancier was grumbling to us about the size of his birds, and on going to his yard we found saucers in all the different pens filled with wheat, oatmeal, etc. It is absurd to expect to keep down size when such an extremely injudicious method of feeding is resorted to.

PREVIOUS ALLIANCES.

Many writers are of opinion that—especially is this mentioned as being the case with a first alliance—when a hen has once been crossed with a bird of another breed she is liable to ever afterwards throw chickens with more or less a trace of this first or former alliance. It is supposed that either the ovum may be partially fertilised by one alliance and the fertilisation completed by another, or that the female's reproductive system having once given birth to offspring having a strongly-marked character, becomes in a degree *moulded* to that character, and tends again to produce it.

We have had the management of too many thousands of birds of different breeds that have been allowed to run together in the off season, and as young birds, to put faith in either the one theory or the other. Our opinion in regard to the first theory is that it is an impossibility when once an egg is fertilised for any subsequent alliance to in the slightest degree alter that alliance. If you cross two breeds together you will always get some chickens that take more after one of the parents than the other chickens do; and simply because certain cases have been quoted where a hen after having had an alliance with a different breed has been placed with a male of her own breed the later chicks produced by her showed more of her breed than they did of the former male, is, in our opinion, a mere coincidence with no proof of the correctness of the theory at all. As we said before, we consider it an impossibility to in anyway alter an alliance when once the egg is fertilised.

Then as to the theory that a hen that has once given birth to chickens with strongly marked characteristics, her reproductive system is liable to become moulded to produce the same again, we would point out that in dealing with a fowl it is a totally different thing to dealing with an animal. In the latter case, the young grows and forms inside the ovum itself, and if it has any strong-marked characteristics, undoubtedly it is liable to mould the ovum to that shape, which would also dispose it to reproduce similar characteristics in the future. But how is that possible with a hen? We utterly fail to grasp its possibility. A hen will lay the same shaped egg when no alliance has taken place, as she will when one is permitted. The chick, moreover, is not matured in her ovum, but entirely separate from it. The yolk is more or less formed in the ovary before fertilisation takes place, and, surely, it cannot be argued that the minute seed of the male (which is not, as many people imagine, that white cord that is seen when an egg is broken) can possibly *mould* the ovum into a given shape.

Our opinion is, that when once a hen has laid the batch of eggs that she was laying, or about to lay, at the time an alliance took place, all effects of that alliance are at an end, and until that batch is laid, absolute dependence that the alliance is at an end cannot be relied on.

Time to Hatch.

The exhibitor should, where possible, endeavour to so mate up his birds that he may have reasonable expectations of obtaining eggs at the time that he most desires them. By this we mean certain hens in some breeds are suitable, when mated to different cocks, for the production of either good pullets or cockerels, and as it is with many breeds an absolute necessity to produce early cockerels in order to win at the early chicken shows, he will select for his cockerel-breeding pens out of these hens those that have passed through an early moult and are reddening up for laying, using the remainder and more backward of them for his pullet pens. He will also, where possible, use either a cockerel for his cockerel-breeding pen, or else a well-preserved one-year-old bird, in order that even in very bad weather the eggs may prove fairly fertile.

Brahmas, Cochins, Langshans, Indian Game, Malays, and such-like heavy breeds should be hatched, as far as cockerels are concerned, as early in the new year as possible for autumn competition, though the pullets and the main stock need not be born before March, and even April.

Lighter breeds—such as Leghorns, Andalusians, and their like—when well reared, will not require more than six months before being in prime exhibition trim, and March and April will do well for the production of both sexes. Still, it is just as well to always produce a few January or February birds for the "Royal" and other similar agricultural shows that are held during the summer months throughout the kingdom. We knew an exhibitor whose reputation as a breeder was maintained for years almost entirely by wins at the "Royal" and summer shows. By the time the Dairy Show came round he was out of the running altogether—his stock was not good enough. It was simply through early hatching he was enabled to win at all, for age at these summer shows is nine-tenths of the battle. Rarely, however, are these early-hatched birds of intrinsic value; in ninety-nine out of a hundred cases will the March or April birds bred from the same pen outstrip them, and, for what we might term our "main crop," these are the two months that we would choose.

Sight Influence.

The influence of sight on the future chicken is, perhaps, a matter that we should make a few remarks about. Most writers maintain that, if birds of a certain colour are placed in close proximity to, and in full view of, other birds of a different colour, the chicks produced from such birds are liable to throw feathers partaking of the colour of those in the adjoining pen. The same remarks that we made under the heading of "previous alliances" apply in a great measure to this case. We do not wish for a moment to imply that we are right in this matter, and all other writers wrong; but when anyone has kept large quantities of fowls for very many years, and frequently under such circumstances that adjoining or opposite pens could easily see each other, and have never had a single case where the effects of sight influence was shown, it is one of those things that can be summed up in the words "seeing is believing." We can understand how sight can influence animals; in fact, we have several times proved this with Dutch rabbits; but how it is possible to influence what we might call "the dormant seed of life" that fertilises the egg, will, we are afraid always remain a mystery to us. Were two different breeds allowed to run together in the same pen, we can conceive it possible. though most improbable, that foul feathers might occur, but not from sight influence.

We once hatched a black Hamburgh that very nearly made us a convert to the theory of sight influence. The pen of Hamburghs were in an adjoining pen to some white Leghorns, and the chicken produced had several large patches of white on its body, especially between the shoulders and on the saddle, and the legs were a very pale bluey white in colour. This one was shortly followed by several more similarly marked. As there was no opening between the two pens, and as the Leghorn cock was never found to have flown over into the Hamburgh pen, there seemed no other way to account for the mystery than by the theory of sight influence. One morning we had occasion to come back through these pens, after feeding them, rather sooner than usual, and to our surprise we found one of the black Hamburgh hens quietly feeding with the white Leghorns. We had no time to stop and shift her at the moment, but as soon as we were disengaged we hurried back to take her out of the Leghorn pen, when lo! she had vanished. We eventually found her apparently serenely happy with an extra full crop of food in her own pen. Next morning we purposely waited handy after feeding them, and as soon as the principal part of the food was consumed in the Hamburgh pen she climbed up the eight-foot wire netting like a cat and dropped down in the Leghorn pen. Believers in sight influence as regards fowls in this particular have still another convert to make.

In our estimation it would be a far more tangible theory to suppose that different-coloured birds running together during the moulting season would be liable to the effects of sight influence; and although we do not put this forward as a proved fact, we are by no means sure that such is not the case at times.

COLOUR-FEEDING.

We have dealt with this subject at some length in Section VI., and wish to here merely add that further experiments in this line have shown that carbonate of iron is equally as serviceable and far less expensive than sacharrated carbonate of iron, and, further, that at the time of giving the colour-feed a little melted fat should also be added to the soft food. We would further like to add here that although we were the first to publish matter on the subject of colour-feeding poultry, we do not consider it to the benefit of the Fancy in general that this practice should be followed. But having practised it for many years ourselves, and knowing as we do that a large number of exhibitors do the same, and further, that at the time of writing no one knows how to detect a colour-fed bird in the show pen from one that is not colour-fed, we thought it only fair to put the amateur, as far as we could, on an equal footing with the "old band," and consider that colour feeding should be allowed by the Poultry Club, anyhow during such time as the means for detecting colour-fed specimens are wanting.

CONCLUDING REMARKS.

In concluding this section we would strongly advise the amateur not to feel downhearted and disgusted, simply because he finds the season's produce does not contain a real good bird. Probably a quiet half-hour spent at his stock birds' pen will show him where the fault lies. Although the cock is a fairly deep-breasted bird, the hens are somewhat deficient in this respect, and in consequence so are the youngsters. Although the stock birds have red faces, yet they were bred from stock that were inclined to go white in face, and the fact that they are being in-bred causes the white face to reappear in the chickens; and so with almost any defect: a little study and careful thought will always reveal the secret of your non-success, and then by a judicious alteration in your breeding stock all difficulties can in time be overcome.

We propose in Section VIII. to treat on diseases of poultry and their cure.

WORKS ON PIGEONS.

Archangel Pigeon, The. By A. A. GOODALL. An interesting manual by one who has made the breed his speciality. Contains Club Standard and Ideal Head. Post free, 10d.

Breeding Record. A complete Stud Register and Summary of Show Results. Pages are also given for Receipts and Expenditure, so that Fanciers can keep a useful Record of their hobby. The book will be found equally good for Pigeon or Cage-Bird Fanciers. Post free, 7d.

Carriers and Barbs. By J. B. BROAD and J FIRTH. Illustrated. An interesting and practical handbook on the Breeding and Management of these Standard Varieties. Post free, paper, 1s. 1d.; cloth, 2s. 3d. (Postal Order.)

Dragoons, About. By W. R. FLETCHER. Every minute detail is dealt with by the Author, as well as notes by well-known specialists in the different colours. Fully illustrated. Post free, paper, 2s. 10d.; cloth, 3s. 10d. (Postal Order.)

Jacobin, The: Its Breeding, Management, and Exhibition. By JOHN WATERS. Illustrated. The fullest and best treatise on the Jacobin yet published. Post free, 1s. 1d. (Postal Order.)

Magpie Pigeon, The. By W. E. COOKE. With particulars as to Breeding and management; the Magpie Club Standard. Ideal in Colour. Many Illustrations. post free, 1s. 2d.

Pigeons, Diseases of: Their Cause and Cure. By W. VALE. Post free, 1s. 2d. (Postal Order.)

Pigeons, Practical Handbooks on. By W. FELLOWES. Diseases of Pigeons; Foot and Feeding Pigeons; Pigeon Lofts and Suitable Fittings; Moulting, Showing, etc., Pigeons; Nesting, Hatching, etc., Pigeons; Selecting and Mating of Pigeons. Each post free, 7d. All the above bound in cloth, 8s.

Pleasures of a Pigeon Fancier. By the Rev. J. LUCAS. With Three Coloured Plates. A delightful book for leisure moments. Cloth, gilt top, post free, 2s. 9d.

Show Homer Pigeon: Its Management and Exhibition. By VICTOR WOODFIELD. Contains plate of Club Ideal and a Chapter on the "Flying Homer" by THOS. WALTON. Post free, paper, 1s. 8d.; cloth, 2s. 3d. (Postal Order.)

Tippler Pigeon, The. For Flying and Exhibition. By ARCHIBALD HEPWORTH. Revised by D. H. WEDGWOOD and additions by T. BEECH, A. STEPHENSON, etc. Post free, 1s. 8d. (Postal Order.)

Tumbler, The Long-Faced. By H. CHILD. Revised thoroughly by Lt.-Col. H. W. BRUNO. Illustrated. Contains valuable information on Long-Faced Tumblers and German Beards. Also Club Standard. Post free, paper, 1s. 2d.; cloth, 2s. 2d. (Postal Order.)

Turbit, The Modern. By H. P. SCATLIFF. Illustrated. Enters fully into details of Breeding, Feeding, Housing, etc. Post free, 1s. 3d. (Postal Order.)

Working Homer, The. By "CLAYFIELD." Giving details for the Management of the Racing Homer throughout the year, and Chapters on Lofts, Feeding, different Types, Diseases, etc. Well Illustrated. Post free, paper, 1s. 9d.; cloth, 2s. 9d. (Postal Order.)

Coloured Plates and Post-cards of different Varieties of Pigeons on Sale. List and prices on application.

HELPFUL WORKS ON CAGE-BIRDS.

British Birds. By the late Dr. BRADBURN. Recently revised by ALLEN SILVER, Junr. The work has been enlarged, is very fully illustrated, and is the standard work on the Management of British Birds in Captivity. Numerous Illustrations besides Coloured Plates. Post free, paper, 1s. 9d.; cloth, 2s. 3d. (Postal Order.)

British Finches. By C. PRIOR. Treats of the best known Finches, their Feeding and Management. Illustrations of the Wild Seeds, etc., recommended. Post free, 1s. 1d. (Postal Order.)

Canary Breeding and Management. Illustrated. By "JEROME." Being Plain Hints for Keeping Canaries of all Kinds in Cages or Aviaries. Post free, 1s. 2d. (Postal Order.)

Canary Management Throughout the Year. By JOHN ROBSON. Gives details of work in the Bird Room for every month in the year. Post free, paper, 1s. 3d.; cloth, 2s. 3d. (Postal Order.)

Our Feathered Pets. By Dr. GREENE. Deals with British Birds which are most usually kept. Post free, 1s. 2d. (Postal Order.)

Pet Bird, A. By H. B. RUTT. A Simple and Practical Guide to the Management of Pet Birds. Post free, 7d.

Other Books on British and Foreign Birds, and on different Varieties of Canaries on sale. List on application.

Those interested in Cage-Birds should see
"Canary and Cage-Bird Life," of which a specimen
copy will be sent on application to this office.

"THE FEATHERED WORLD," 9, ARUNDEL STREET, STRAND, LONDON, W.C.